NATIONAL GEOGRAPHIC KiDS

美国国家地理 双语阅读

Tigers
老虎

第三级

懿海文化 编著

马鸣 译

外语教学与研究出版社
FOREIGN LANGUAGE TEACHING AND RESEARCH PRESS
北京 BEIJING

京权图字：01-2021-5130

图书在版编目（CIP）数据

老虎：英文、汉文／懿海文化编著；马鸣译. —— 北京：外语教学与研究出版社，2021.11（2023.8 重印）
（美国国家地理双语阅读. 第三级）
书名原文：Tigers
ISBN 978-7-5213-3147-9

I. ①老… II. ①懿… ②马… III. ①虎－少儿读物－英、汉 IV. ①Q959.838-49

中国版本图书馆 CIP 数据核字 (2021) 第 236728 号

出 版 人　王　芳
策划编辑　许海峰　刘秀玲　姚　璐
责任编辑　姚　璐
责任校对　华　蕾
装帧设计　许　岚
出版发行　外语教学与研究出版社
社　　址　北京市西三环北路 19 号（100089）
网　　址　https://www.fltrp.com
印　　刷　天津海顺印业包装有限公司
开　　本　650×980　1/16
印　　张　37.5
版　　次　2022 年 3 月第 1 版 2023 年 8 月第 4 次印刷
书　　号　ISBN 978-7-5213-3147-9
定　　价　188.00 元（全 15 册）

如有图书采购需求，图书内容或印刷装订等问题，侵权、盗版书籍等线索，请拨打以下电话或关注官方服务号：
客服电话：400 898 7008
官方服务号：微信搜索并关注公众号"外研社官方服务号"
外研社购书网址：https://fltrp.tmall.com

物料号：331470001

Table of Contents

PURR-fectly Big Cats

Bengal Tiger

Tigers are big and beautiful animals. They are strong and powerful, too. Tigers are the biggest cats in the world.

A Tiger's Home

Tigers live in the forest. They spend a lot of time in the water, too. They live in hot places like Indonesia. They live in cold places like Russia.

Indochinese Tiger

Siberian Tiger

Tigers that live in cold places are bigger than other tigers. They also have thicker fur to keep them warm.

Built for Hunting

Tigers are fierce hunters. Their bodies are built for catching prey.

Coat

A tiger's stripes camouflage it in tall grass and dry leaves. Its prey may not see the tiger until it's too late.

Teeth

Four large teeth help tigers kill prey quickly.

Eyes

A tiger's terrific eyesight helps it hunt at night.

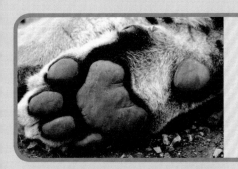

Paws

Big paws with soft pads help a tiger walk quietly. Sharp claws hook into prey and don't let go.

Tail

A long tail helps a tiger keep its balance when moving quickly.

Back legs

Big muscles help a tiger dash or leap at its prey.

Tiger Terms

CAMOUFLAGE: An animal's natural color or shape that helps it hide from other animals

PREY: An animal that is killed and eaten by another animal

Meat Eater

Tigers are carnivores animals that eat meat. Their favorite foods are large, hooved animals such as buffalo, deer, and wild pigs.

A hungry tiger can chow down 80 pounds of meat in one meal. That's about 320 hamburgers!

Tiger Turf

Besides hunting, tigers spend a lot of time marking their territory. They are not good at sharing!

Tigers make long scratch marks on trees. They also rub their faces on trees and leave smelly scents. This tells other tigers to stay away.

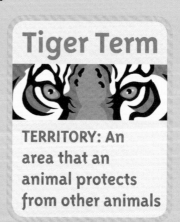

Tiger Term

TERRITORY: An area that an animal protects from other animals

Siberian Tiger

Cool Cat Facts

Check out these neat facts about tigers.

No two tigers have exactly the same stripes.

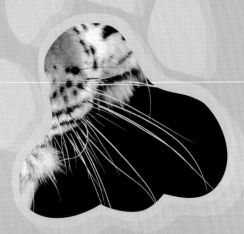

Whiskers help a tiger feel its way in the dark.

Large paws keep Siberian tigers from sinking in deep snow.

Tigers can live in temperatures as low as -40 degrees Fahrenheit.

Tigers are great swimmers. They are never far from water.

A tiger's front teeth are three inches long.

Tigers have much better hearing than humans.

Cubs

A female tiger usually has two or three cubs at one time. The cubs weigh about four pounds at birth.

The mother raises the cubs by herself. Male and female tigers come together only to have cubs. Otherwise, adult tigers live alone.

Bengal Tigers

A mother Bengal tiger nurses her young.

The cubs drink their mother's milk. After three or four months, they start to eat meat.

Tiger cubs play games. They chase, leap, and pounce. They are learning how to be good hunters. When they are two years old, young tigers leave their family to find their own territories.

Tiger Talents

Tigers are full-grown when they leave their families. They are big, heavy cats, but they can climb trees and jump great distances.

Bengal Tiger

In fact, tigers can leap as far as 30 feet. That's as long as five adult men lying head to toe!

And, unlike house cats, tigers are good swimmers. They like to cool off in rivers and pools.

Sumatran Tiger

The White Tiger

The white Bengal tiger is very rare. It can't grow orange fur. Its white coat, brown stripes, and icy blue eyes are quite a sight.

You won't find a white Bengal tiger in the wild. But you might be able to see one in a zoo.

Q What's striped and bouncy?

A A tiger on a pogo stick!

White Bengal Tiger

Tigers in Trouble

Tigers are endangered. About 100 years ago, there were 100,000 tigers in the wild. Today there are less than 4,000.

There are five different kinds of tigers today. They are the Bengal (BEN-gol), South China, Indochinese (in-doh-chi-NEEZ), Suma-tran (soo-MAH-truhn), and Siberian (si-BEER-ee-uhn) tiger. Three other kinds of tigers have already become extinct.

Tiger Terms

ENDANGERED: At risk of dying out

EXTINCT: A type of plant or animal no longer living

Why are tigers disappearing?

Tigers are losing their habitat. People cut down trees. Tigers live and find food in the forests. When forests disappear, so do tigers.

People also kill tigers for their body parts. Their skins are used for rugs.

Killing tigers is against the law. But it still happens today.

Tiger Term

HABITAT: The place where a plant or animal naturally lives

South China Tiger

27

Helping Tigers

Bengal Tigers

Though tigers are in trouble, there is good news. New forest areas for tigers have been found. Also, people are planting trees where forests have been cut down.

You can help, too. Tell your family and friends about what you've learned. We can all work together to keep tigers on our planet!

Stump Your Parents

Can your parents answer these questions about tigers? You might know more than they do!

Answers are at the bottom of page 31.

1

How do tigers spend their time?

A. Howling
B. Hunting and marking their territory
C. Sharing their territory
D. Eating plants

2

What is special about a tiger's paws?

A. They are small but powerful.
B. They don't sink in deep snow.
C. They have three toes.
D. They make noise when walking.

3

How do tigers live?

A. Alone, except when a mother is raising her young
B. In groups of three to four
C. In groups of five to eight
D. With all their friends

4

What's important about a tiger's striped coat?

A. It sticks out.
B. Its pattern is the same as other tigers'.
C. It comes in many colors.
D. It camouflages the tiger.

5

What do tigers like to eat?

A. Fruits and berries
B. Insects
C. Meat—and lots of it!
D. People

6

Where do tigers like to live?

A. In the desert
B. Near water
C. In the mountains
D. On the savanna

7

What is a baby tiger called?

A. A pup
B. A kitten
C. A cub
D. A gosling

Glossary

CAMOUFLAGE: An animal's natural color or shape that helps it hide from other animals

ENDANGERED: At risk of dying out

EXTINCT: A type of plant or animal no longer living

HABITAT: The place where a plant or animal naturally lives

PREY: An animal that is killed and eaten by another animal

TERRITORY: An area that an animal protects from other animals

▶ 第 4—5 页

完美的大型猫科动物

孟加拉虎

老虎是体形大、外形漂亮的动物。它们也强壮有力。老虎是世界上最大的猫科动物。

▶ 第 6—7 页

老虎的家

老虎生活在森林里。它们也会在水中待很长时间。它们生活在印度尼西亚这样的炎热地区。它们也生活在俄罗斯这样的寒冷地区。

印度支那虎

西伯利亚虎

生活在寒冷地区的老虎要比其他老虎更大。它们的毛皮也更厚，可以维持自身的体温。

为捕猎而生

老虎是凶猛的猎手。它们的身体就是为捕猎而生的。

毛

老虎身上的条纹让它能隐身在高高的草丛和干枯的树叶中。当猎物发现老虎时，已经太迟了。

牙齿

四颗大牙让老虎能迅速咬死猎物。

眼睛

老虎的视力非常好，让它能在夜间捕猎。

脚掌

长着柔软肉垫的大脚掌使老虎能悄悄地行走。锋利的爪子一旦抓住猎物，就不会让它逃脱。

尾巴

长长的尾巴让老虎能在快速移动时保持平衡。

后腿

大块的肌肉使老虎能猛冲或跳向猎物。

老虎小词典

伪装：动物天然的颜色或形状，帮助它不被别的动物发现

猎物：被另一只动物杀死并吃掉的动物

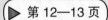 第 10—11 页

肉食性动物

　　老虎是吃肉的肉食性动物。它们最喜欢的食物是高大的有蹄动物，比如水牛、鹿和野猪。

　　一只饥饿的老虎一顿能吃掉 80 磅（约 36.29 千克）肉。那相当于大约 320 个汉堡包！

▶ 第 12—13 页

老虎的地盘

　　除了捕猎，老虎还会花很多时间来标记自己的领地。它们不擅长分享！

　　老虎在树上留下长长的划痕。它们还用脸在树上摩擦，留下难闻的臭迹。这告诉别的老虎不要靠近。

西伯利亚虎

老虎小词典

领地：一个区域，动物保护它不被别的动物夺走

▶ 第 14—15 页

酷猫酷事

　　来查一查这些关于老虎的事实吧。

胡须帮助老虎在黑暗中感知方向。

任意两只老虎的条纹都不一样。

大大的脚掌使西伯利亚虎不会陷入厚厚的积雪。

老虎可以在气温低至零下40华氏度（等于零下40摄氏度）的环境中生存。

老虎是游泳高手，它们从不会远离水。

老虎的犬齿有3英寸（约7.62厘米）长。

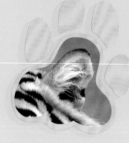
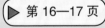
老虎的听觉比人类灵敏多了。

▶ 第 16—17 页

虎崽

　　母虎通常一胎生两只或三只虎崽。虎崽出生时大约 4 磅（约 1.81 千克）重。

　　母虎独自抚养虎崽。公虎和母虎只有在孕育虎崽时才在一起。除此之外，成年老虎都独自生活。

孟加拉虎

一只孟加拉虎妈妈在给孩子喂奶。

虎崽吃妈妈的奶水。三四个月之后，它们开始吃肉。

虎崽玩游戏。它们追逐、跳跃、扑袭。它们在学习如何成为优秀的猎手。两岁大时，小老虎会离开家，去寻找自己的领地。

老虎的天赋

老虎离开家的时候，它们已经发育成熟了。它们是又大又重的猫科动物，但它们能爬树，跳得也很远。

孟加拉虎

苏门答腊虎

事实上，老虎可以跳 30 英尺（约 9.14 米）远。那相当于 5 个成年男人头脚相连的长度！

与家猫不同，老虎还是游泳好手。它们喜欢在河里和池塘里消暑。

白虎

孟加拉白虎非常罕见。它长不出橙黄色的毛皮。它那白色的毛、棕色的条纹和冰蓝色的眼睛是亮丽的风景。

在野外你是找不到孟加拉白虎的。但在动物园里你可能会看到一只。

孟加拉白虎

困境中的老虎

老虎是濒危动物。大约 100 年前，有 100,000 只野生老虎。现在有不到 4,000 只。

老虎现存 5 个亚种，它们是孟加拉虎、华南虎、印度支那虎、苏门答腊虎和西伯利亚虎。另外 3 个老虎亚种已经灭绝了。

老虎小词典

濒危：有灭绝的危险

灭绝：一种植物或动物不再存活

▶ 第 26—27 页

老虎为什么会消失呢？

老虎正在失去它们的栖息地。人们砍伐树木。老虎在森林里生活和觅食。当森林消失时，老虎也就消失了。

人类还会为了老虎的身体部位而杀害它们。虎皮被用来做地毯。

杀害老虎是违法的。但现在这依然存在。

老虎小词典

栖息地：植物或动物天然生长的地方

华南虎

▶ 第 28—29 页

救救老虎

孟加拉虎

虽然老虎身处困境，但还是有好消息。人们为老虎找到了新的林区。此外，人们正在被砍伐的森林里种树。

你也可以尽一份力。告诉你的家人和朋友你学到了什么。我们可以一起努力，让老虎留在地球上！

挑战爸爸妈妈

你的爸爸妈妈能回答这些有关老虎的问题吗？你可能比他们知道的还多呢！答案在第 31 页下方。

1 老虎平时怎么打发时间？
A. 咆哮　　　　B. 捕猎和标记领地
C. 分享领地　　D. 吃植物

2 老虎的脚掌有何独特之处？
A. 它们小而有力。　　B. 它们不会陷到厚厚的积雪里。
C. 它们长着三个脚趾。　D. 走路时，它们会发出声音。

3 老虎怎样生活？
A. 独居，除了妈妈养育孩子的时候
B. 三到四只群居　　C. 五到八只群居
D. 与所有的朋友一起

4 老虎带有条纹的毛有什么重要作用呢？
A. 它很醒目。　　　　B. 它的模样与别的老虎相同。
C. 它有很多种颜色。　D. 它帮助老虎伪装。

5 老虎喜欢吃什么？
A. 水果和浆果　　　　B. 昆虫
C. 肉——很多肉！　　D. 人

6 老虎喜欢在哪里生活？
A. 在沙漠里　　　B. 在水附近
C. 在大山里　　　D. 在稀树草原上

7 虎宝宝叫什么？
A. 幼犬　　　　B. 小猫
C. 虎崽　　　　D. 幼鹅

▶ 第32页

词汇表

伪装：动物天然的颜色或形状，帮助它不被别的动物发现

濒危：有灭绝的危险

灭绝：一种植物或动物不再存活

栖息地：植物或动物天然生长的地方

猎物：被另一只动物杀死并吃掉的动物

领地：一个区域，动物保护它不被别的动物夺走